TWISTED METAL

ISBN 978-1-946886-39-2

Library of Congress Control Number: 2022917647

Front cover photo by Bruce Borich

Back cover photo by Gustavo Rivera

Published by

Middle River Press
Oakland Park, Florida
middleriverpress.com

First Printing

Printed in the USA

TWISTED METAL

JON MALEK

MIDDLE RIVER PRESS

Foreword

You are about to embark on an amazing journey, as you will now experience the life of an aircraft accident investigator. This book is going to take you on many adventures – including major fatal accident investigations, criminal investigations such as murder and suicide, strange crashes that to this day remain unexplained, and just about anything else you can imagine. This book literally puts you in the driver's seat of what this type of life is like. So please sit back, relax, fasten your seatbelt, secure your tray table, and put your seat in the full upright and locked position. Enjoy!

Dedication

My father was my biggest supporter of my initial flight training. He also encouraged me the most, especially later in life when considering getting a master's degree. If it was not for him, I may not have either. That's a scary thought.

While my mother was a bit more cautious about my flight training, she was the one who drove her only son (and only child) to the airport on many hot South Florida summer days to watch me hop into a two-seater airplane and learn how to fly with my instructor by my side. Keep in mind – I was fourteen years old.

I can't thank my parents enough for all of the support they gave me – not just in my career path, but in life. Thus, I am honored to dedicate this book to them.

Thank you, Mom, for always being there and even helping me with the writing of this book! Just today I called you for advice, and you were on point. Dad, I know you are cheering me on in heaven. Miss ya!

I would also like to dedicate this book to those who have lost their lives to aviation, especially to the ones I knew personally (several of whom while writing this book).

Lastly, it would not be right if I did not also dedicate this to my fellow aircraft accident investigators – the damn few.

How I Became an Aircraft Accident Investigator

It was an early morning flight. A little stormy, but not bad by any means. We had what seemed to be a very long taxi out to the runway. As we taxied onto the runway, two audible dings advised the flight attendants that we were moments from takeoff.

After a brief moment on the runway, the engines roared to life and the aircraft started moving. As we accelerated faster down the runway, the nose gently lifted off and we were flying. Soon thereafter you could hear the landing gear retract. And about a minute after that the flaps started coming up.

I was in the middle of the plane at an exit row aisle seat, which is usually my favorite. There were a few other passengers around me for this short flight from New York to Florida. Some were already asleep before takeoff. One passenger near me was sipping on a cup of coffee he bought in the terminal before we boarded.

All of a sudden, the plane started to shake and shimmy violently. The engines made different noises than usual. Before we knew it, the nose pitched up and then dropped suddenly. No roller coaster I had ever been on had felt this violent. Everyone was awake and I could hear some passengers gasping.

Yet again, the nose pitched up, this time more violently. And then the nose seemed to dive over as it felt like we were now going straight down. This was shortly after takeoff, so I knew that was not good. My fellow passengers and I were thinking, we may crash.

A million things go through your mind when you are in this particular type of situation. Is there anything I can do? Why is this happening? Is there any way I can get one last message off to my loved ones? Most of us just wanted this plane back on the ground safely. But that was not gonna happen.

As we came out of the clouds, the nose was still pointed almost straight downward, and I could see the ground coming out at us quickly. I took a deep breath, and then I woke up. Thankfully, it was just a dream.

Even though I was a young teenager when I had this dream, I still wanted to know why I had it and what it meant. Did it really start my fascination with aircraft accident investigation?

Intro Flight $20

At the age of fourteen, one of the things I hated most in my life was walking through the grocery store with my mom. There are so many other things I wanted to do at fourteen years old. Ride my bike, play with my friends, launch model rockets, ride my all-terrain vehicle, horseback riding, going to the beach. You name it. Anything but going to the boring ass grocery store. So I would grab the community newspaper, which thankfully was free (we didn't have much money back then) and peruse through it as I walked up and down the mundane aisles.

And sure enough, this time something really caught my attention. If you would've told me that my entire life was about to change forever, I wouldn't have believed you. I saw an advertisement for an introductory flight for $20! I asked Mom if I could do it and she said absolutely not. So what did I do? What any good son would do in that situation. I asked dear ole Dad when I got home. He said yes, as he had a childhood friend who was now a pilot for United Airlines. I think he was happy that I was interested in something that involves math and science. Growing up, I was a good kid and did very well in school. I just wasn't really applying it toward anything important yet. That was about to completely change.

My dad made the arrangements for my first introductory flight, and the next weekend it was game on! We drove to a local municipal airport and I went right to the hangar where the airplane was kept. This was in the early '90s, before the days of heavy airport security, multiple fences with electronic gates, cameras, etc. We literally just drove around to the hangar. It was big and blue, and full of small aircraft. Now the advertisement had a picture of an airplane and a helicopter. The helicopter was down for maintenance, allegedly. So, it was the airplane for me. We walked out to an aircraft that was smaller than my dad's car. It was a Cessna 150. Two seats and very small. But I didn't care. At that point it was one of the most interesting things I had ever seen in my entire life. And I was about to go up in it.

My flight instructor's name was Chris. He was likely in his mid-20s, which back then I considered quite old. We got into the airplane, which smelled like aviation gasoline, leather, and old metal. Our shoulders always touched, as this was a very small cockpit. I sat in the captain's seat on the left. My new flight instructor sat on my right. We taxied out to the runway, and he told me I was going to do the takeoff. I was very excited, to say the least.

I had a flight plan in my mind and even drew it out the night before. I wanted to take off and go to the beach, past the pier, then around the lighthouse and then back. So, I showed this to the instructor, which I drew with a #2 pencil on an 8 1/2 x 11 sheet of looseleaf Trapper Keeper paper. He made some small edits due to airspace, but he liked it.

Back to the takeoff. He lined up the airplane on the runway center line and told me to grab the controls, push the throttle forward, and gently pull back on the controls when he told me to. This was much easier than I thought it would be. So yes, on that day, I flew a plane before I could legally drive a car.

OK, I was hooked. It was love at first flight. After landing, Chris asked me if I wanted to start a logbook. I asked him what a logbook was, and he said it was basically a ledger to keep track of all my flights and one day become a professional pilot to get paid for doing this. Man did that sound great.

Moolah anyone? Houston. We have a problem.

OK, I was very interested in this hobby. However, it is a VERY expensive hobby. And as I would later learn, becomes increasingly more expensive, exponentially. The introductory half-hour flight was $20. Each full flight hour thereafter is $29. The flight instructor is another $10 per hour. You need a minimum of 40 hours of flight to become a private pilot. This does not include approximately 100 hours of ground training, books and materials, etc.

I grew up in a middle-class home, in a middle-class neighborhood in the 1980s. We had a lot and were not considered poor; however, $20 was a lot of money back then. Most things we bought cost way less than that.

My parents were graciously willing to contribute, but in no way could they pay for the entire training. While I was an entrepreneur in spirit and made some money here and there, there would never be enough.

Would you believe that my dog made me a pilot?

Around the time I started becoming interested in flight training, I received a call from the local vet about breeding my dog. Growing up, we got a male English springer spaniel and we named him Fred. Since I was unfamiliar with this process, I asked more about it. After breeding your dog, you either get the pick of the litter or the value of the pick of the litter. Fred was my dog and my very best friend. That dynamic was never going to change, so I was much more interested in the value of the pick of the litter versus getting another dog. So, guess where that money went? Yes, almost all of it went toward flight training. Also started working a couple odd jobs and my parents contributed, when they could. Needless to say, Fred was also happy about his part of the deal!

Over the years I flew more and more often and accumulated a good amount of flight time. Around my sixteenth birthday, it was time for my first solo flight. Well, I wore a shirt with a picture of Fred ironed into it, in his honor.

In school I was quite interested in business and almost went that route in life. However, my local community college offered an aviation program and I enjoyed it. So, I dropped out of high school and started the aviation program at the local community college at 16. Back then they didn't really have a college bridge experience program in high school. So, I guess I kind of started my own pathway.

While I did well in that program at community college, it wasn't until I started pumping gas in airplanes at my local airport that I really knew aviation was the career path for me. So, I transferred into Embry-Riddle Aeronautical University to train to become a professional pilot. Community college was great. But this place was amazing. Everyone there had the same love that I had. Everyone there was bitten by the same bug that bit me. The aviation bug.

As one can imagine, flight school was amazing and challenging at the same time. I perhaps did not quite realize back then that actual events which occurred during flight school were preparing me for my career.

For example, my first experience with aircraft theft was actually when my secondary flight training instructor got drunk and stole a flight school airplane. After three takeoffs and landings, tower asked him if this would be his final landing of the night and he responded yes it sure will be. He flew the airplane straight down into the runway and it exploded upon impact. A suicide note was later found in his apartment in which he claimed that he was tired of life. I had seen him walking out of a bar that night with a soulless look on his face that I will never forget. He later jumped the airport fence and stole an airplane. I was one of the last people to see him alive. Little did I know at the time, this would not be my last experience with aircraft theft or suicide.

My guidance counselor and I felt it would be good to diversify my education, so I picked up minors in safety, psychology, and flight. A few classes focused on aircraft accident investigation. Holy crap, did I love this stuff. So, I firmly decided there and then, that I would be a professional pilot and an aircraft accident investigator for the rest of my life. I even changed my major to further specialize in this field. This was not easy by any means. Most days I went to school all day and flew all night. In three short years, I had a degree in hand and a commercial pilot certificate. Look out world, here I come.

Baptism of Fire

Before I even graduated, I had many job applications out there with my name on them seeking a chance at doing what I knew I would really love. Luckily, many seemed interested in me. Several offers came in from several organizations. And just like that, I landed a desirable position based in New York. While this was a major culture change for me as I was originally from the South, I was way too excited to let anything faze me. Within my first month, I was dispatched with another junior investigator to investigate my first on-scene fatal plane crash. Accidents in my world have many names, or ways to describe them. Common terms include plane crash, accident, incident, fatality, loss, fatal, etc. Now back to this loss. I flew us in a Cessna from New York to Delaware, then we rented a car and drove to the crash site. The body of the pilot, and sole occupant, had already been removed by the time we got there. As I continued my career, that generally was the norm – which was fine with me. It's my job to investigate the plane crash, not look at dead people – although that would happen from time to time. A twisted and mangled pile of metal lay in a cornfield in what seemingly was the middle of nowhere. It appeared the pilot was flying low and slow, and eventually stalled the airplane before crashing. Unfortunately, this type of scenario will be an almost common occurrence throughout my career as an aviation accident investigator. People think of the word "stall" and relate it to an engine. This is not that kind of stall. When an aircraft goes so slowly it aerodynamically no longer flies, it is called a stall.

Part of being an aircraft accident investigator, we look at man or woman, machine, and environment. We're trying to determine what did not happen and see what we are left with. For example, if there was no pre-crash or post-crash fire, we would like to make sure there was enough fuel onboard the airplane for the aircraft engine(s) to run. We look at the marks on the ground to see if the prop was spinning or not when the aircraft impacted the ground. This will tell us if the engine was producing power or not upon impact. We also interview witnesses and check for any surveillance camera footage. Lastly, we usually will check the weather and any other external environmental impact potential. These are some very basic and general steps we follow in most accidents.

We called my boss and reported our findings. We then made our way back to the airport and flew home, landing by sunset. I spent most of the next couple of days writing up the accident report for the file. And just like that, I was officially an aircraft accident investigator.

My flight instructor, Bob, who got me through the private pilot check-ride, was amazed when I told him what I wanted to do with my career. He always thought of it as a very dark occupation with lots of smoldering and burnt stenches. He was tainted by another investigator who could not handle the position and eventually sought another career. For me, this was not the case. I had a job to do, and always focused on doing it to the best of my ability. Tragedy comes with the territory – like innocent people on the ground being killed and injured. Thankfully that is very rare. Most of my investigations do not even include injuries.

When I started as an aircraft accident investigator, things got busy real quick. I was usually traveling out to crash sites several times per week. Many times, I would be dispatched to another plane crash while I was on scene at the first one I was dispatched to. Back then there was not a huge emphasis on safety, at least from my perspective. And that bothered me. I wanted to do something about it, and eventually I made a contribution. More about that later…

Flying a small plane around the Northeast was usually incredible. I especially loved flying in the fall when the leaves were changing. It was like looking down on rolling hills of fire. I will never forget it. However, it also came with many challenges. I remember flying with one of the senior investigators to evaluate damage to a jet during the wintertime and he warned me to watch out for black ice. Being young and from the South, I had no idea what he was even talking about. Plus, winter turbulence from cold fronts was sometimes like being in a popcorn machine. There were a few flights where I said if I get this plane on the ground safely, I will never fly again. Thankfully, there were only a few of those occasions.

Life in NYC

Living in New York was a major culture shock. For starters, my office in Manhattan faced the NYU girls' dorm. If you can imagine it, I have seen it. Even just being on the subway was wild, from seeing the suits get off at Wall Street to the freaks getting off in the Village. Back then, there were Hallmark retail stores. That's where we would go to buy a cake and get a card if it was someone's birthday that we worked with. Well, one time when doing a birthday cake run my co-worker and I saw some girl squatting and masturbating in the corner of a Hallmark retail store behind a greeting card display. This is something you just can't make up.

One snowy day a fellow commuter dropped his briefcase onto the train tracks. When he climbed down to pick them up, he slipped, fell, and landed on one of the tracks which ended up going into the middle of his back. Sure enough, he fractured it and we never saw him again. For those of you who have never lived in New York or a major commuting city, you can't imagine the hours every week you spend on a train or bus, and hours you spend just waiting for one. Sometimes that alone was 40 hours per week. I called myself and my fellow commuters train watchers. One good thing that comes out of it is you spend years with the same people and forge amazing lifelong relationships!

Nothing could prepare me for this...

It was a beautiful Tuesday morning in September. One of the best days I had seen since I moved to New York City 18 months prior. My office in downtown Manhattan did face the NYU dorms. However, it also faced the World Trade Center towers. Around 8:45 that morning I saw our team leader running and asking if anyone had a radio. When I turn around in my chair to see what the problem was, I noticed smoke coming out of one of the towers. Immediately, I thought it was a bomb. Smoke was billowing upward from the middle of the building. It appeared a bomb had gone off. But soon thereafter, the second plane hit, and we knew for sure what the problem was. We were under attack.

The accident investigator in me kicked in and I started taking pictures of everything. Most were pretty calm until the first building collapse. Then all hell broke loose. We were on the 15th floor and decided to start an evacuation down the stairs. When we got to the bottom, we saw a lot of people with tons of debris all over them, some injured. So, venturing out of our building at that point did not seem like a good idea after all. Eventually, we went back upstairs to our offices on the 15th floor and waited it out. Around 1 p.m., we started heading home – on foot.

The horrible events of September 11, 2001, did not have a tremendous effect on me. But I do firmly recall hearing sirens in my head for days afterward. The reason being, as I surmise, is because I heard sirens nearly all day on the day of the loss. My dad was driving home a couple of days later from his vacation in Vermont. I hitched a ride with him and stayed home in the South for about a week before returning to New York City. Things got back to normal pretty quickly, which I was just fine with.

OK, here's a pop quiz for you. I walked home on September 11, 2001. I also involuntarily had to walk home on August 14, 2003. Why? If you guessed that was the day of the black-out, you would be correct! That was also a horrible experience. Had to walk home in the dark while it was very hot. I do have to say though, for every bad experience in New York there were so many more great experiences. Amazing restaurants, incredible shows, concerts, fantastic sites; and almost always something to do. I would often have friends and family visit, and we would always do it up like tourists. I am forever grateful for all the experiences I had in the 11 years I lived and worked in New York.

My First Theft

As I was still considered a junior investigator, for the most part I was being assigned smaller cases. Minor aircraft damage investigations, single fatal accidents, and an occasional multi-fatal accident. So when this next case came in, I'm sure my higher-ups viewed it as something very small and simple. Little did I know, it would practically be career-changing for me.

It was a normal day at the office just like any other. Early that morning I was assigned an aircraft theft claim out of Connecticut. It was an older plane. A 1975 Cessna 182. It has four seats, 200 hp single prop engine, and a 600- to 800-mile range. While this seemed as simple as a basic auto theft, I was very soon going to learn it was not.

The initial circumstances were suspicious. I spoke with the owner of the airplane, and he told me he lent the aircraft out to a friend of his who moved it to an airport about an hour's flight away for some meetings. So immediately both the owner and the pilot were suspects. Their stories seemed way too unique and rehearsed – and just did not add up. I met with them both a couple times, and their stories never wavered. So, the investigation continued.

Given the nature of the case, a state special investigator was assigned. She was very energetic and seemed to have a unique interest in this case. I too felt the same way. We learned that a local police officer kept a log of all aircraft at this airport around 9 a.m. each day. The aircraft was last seen around 7 a.m. the day of the theft. It was not on the police log at 9 a.m. So we know it was stolen between 7 and 9 a.m. I immediately contacted an air traffic control radar expert. He was an older gentleman and probably very bored as he was nearing the end of his career. He offered to assist us at no charge. So we went to archived radar and tracked several aircraft that left the airport between 7 and 9 a.m. on the date of the theft. The FBI also assisted us.

One aircraft went to the shoreline and flew very low, according to radar we were tracking from several days prior. We were convinced this was the plane. Why else would somebody be flying an airplane that low, that early in the morning? Well, we later found out it was a banner-tower airplane. Another aircraft flew south down toward Maryland. We were lucky and found that airplane, which was not the one we were looking for. During this time, we also did offer a reward for anyone who saw the airplane and aided in the return of same. So you can imagine I was getting phone calls almost every hour with people claiming they saw the airplane. None of them panned out. This was the time before cellphone cameras, and nobody got a good picture of it.

Several days had passed and the case was starting to go cold. We had tracked almost every airplane that left that morning without success. There was just one left. One plane had gone north. It seemed like a normal route that many aircraft would take up and down the East Coast. But this one seemed a little different, as it landed in Canada. We were not able to locate a flight plan. And that was a little suspicious. We called several airports in Canada and none of them recalled seeing it. Once again, the case was starting to go cold.

We started tracking flights from Canada back into the United States. This took a couple days, but we found one that went into Illinois. The problem was there were many airports nearby where we lost contact with the aircraft. We also had no idea if this was the aircraft or not. We were also still interviewing local possible suspects. No luck though.

After asking several local airports in Illinois to check their fuel receipts for this missing aircraft registration number, kind of like the license plate on your car, sure enough we found one. We searched the entire airport, and it was not there. However, this gave us hope. Our first actual break in the case.

We checked another nearby airport that did not have fueling capabilities and found nothing. One last airport to check. We spoke with the airport manager there and he said a new plane had just registered with the airport and got the proper insurance documents. While we were not hopeful this was it, we still double-checked. As pure luck would have it, this was it! But what do we do? The aircraft was in a hangar and the thief/pilot should be arrested.

After searching the aircraft registration documents with the airport and the insurance application information, we quickly learned who the likely thief was. The special investigations unit from Connecticut coordinated with the local police in Illinois, and one early cold morning, the arrest was made. The logbooks to the aircraft were found in the suspect's home, as well as other documents that belong with the aircraft. We got our man.

But why?

Why did this guy steal this airplane? Did he know the owner or the pilot that moved the airplane when it was lent to him? Why did it go from Connecticut to Illinois? Most of this would not be learned until the criminal trial. The suspect, who decided to represent himself in court, claimed the aircraft was a repossession. He seemed to have fabricated documents supporting the repossession. However, he was not a repo man himself nor did the documents make any sense. We later found out he was incarcerated in Alaska for lumber theft and while in prison learned how to steal and make it look like a repossession. Well, the scheme did not convince the judge and he was found guilty. What do you think the sentence was for aircraft theft across state lines? If you were to guess 20 years, or 10 years, five years, or two years, you'd be wrong. The sentence was 18 months. That's it for a repeat offender who created a wild multi-country and multi-state federal goose chase.

In the end, the aircraft was thoroughly inspected and returned to the aircraft owner. He was very pleased to get it back. Selfishly, I was also thrilled to be such a key part of this investigation and conviction. The FBI recognized my contributions to the case and thereafter designated me a Key-Man for them. Unfortunately, this would not be the last time I worked with the FBI. More to come on that later…

The One That Got Away

When I first moved to New York I lived in a pilot crash pad with a good friend of mine. A crash pad is a shared house with mostly airline pilots that stay there the night before they fly a trip but live somewhere else. I was lucky to meet some very interesting and accomplished pilots, many of whom I am still friends with to this day. I also joined a couple pilot organizations in Manhattan and networked like crazy.

One day I met a couple pilots at a social event, and we decided to grab dinner afterward. Mike and Jeannie were new airline pilots, and new to NYC. I was happy to show them around. Mike actually played football for Notre Dame before becoming a pilot and this was Jeannie's second airline she was now working for. They were both very interesting people. Super sweet as well.

There was something special I immediately noticed about Jeannie. She was a little soft-spoken, very smart, gorgeous, already had an interesting career, and was embarking on her final career move. Do you know that movie *There's Something About Mary*? Well, she is the real Cameron Diaz from that movie. Everyone she meets just absolutely adores her. Needless to say, she had my interest.

While I hung out with many pilots I knew on a regular basis, I tended to hang out with Jeannie more than the others. What was also amazing is the fact that most of the times we met up were spontaneous. Almost none of our outings were planned. She was on reserve for a few years before becoming a line holding captain. So, what being on reserve means is you must be at your domicile/base on a two-hour maximum callout most of the time. Once you are senior enough to hold a line, you then know the majority of your monthly schedule. Until then, you usually have to spend many hours and days at the crash pad ready for them to call you on a moment's notice.

I loved that Jeannie was on reserve and in New York with nothing to do most nights. Her crash pad was near mine. We must have gone out over 100 times over several years. It could be to dinner, a movie, flying, listening to local bands, whatever. We always had a blast doing whatever we wanted, and we were the best of friends. Her soft-spoken way and occasional shyness seemed to always melt away whenever we were together. We would end almost every outing with a walk in the park holding hands in the moonlight, a slow dance, and a kiss good night. Little does she know that she danced away with my heart, and I haven't seen it since, even to this day.

I wish I didn't know.

Over the years we drifted apart. She got a line schedule as a captain, and I did not see her as often. She then changed her domicile/base closer to her home, and I saw her even less after that. According to a mutual friend, eventually she got married and had kids, and that was pretty much it. All those times we went out I was so in the moment that I never worried about the future. There could've been a huge crowd in Grand Central Station with thousands of people. If I was with Jeannie, there was nobody else around. She had all of me and I was happy to give her that.

If you would've told me then she would just be a chapter in my book that I would write one day, I would not have believed you. But I am very thankful for all the time we spent together. Truly one of the sweetest people I have ever met. She was the queen of my nights. Really miss her.

The Godfather

In college I had an economics professor who in a previous career was an executive for a tobacco company in Manhattan. He would share some pretty crazy stories with us. One of which was a vice president in his company that seemed to be more interested in the ladies than his career. He would have dates for lunch both with people inside and outside of the organization he worked for. Needless to say, they were some long lunches!

This seemed a little far-fetched and hard for me to believe as I was always taught that work comes first. My parents were a bit old-fashioned that way. I grew up as a good kid and always did well in school. While I really love dating, I spent most of my years in long-term relationships; it just wasn't in me to think about messing around with girls on my lunch break from work. So, you can imagine that to me this story seemed a bit embellished.

Remember that other junior investigator I went to my first plane crash with? We nicknamed him The Godfather. He was a very laid-back guy and was in a long-term relationship when I met him. Apparently, that girl screwed him over, so it was game on for him thereafter. From then on, he always had multiple girls in play. Both in and out of the office. I couldn't believe it; I was actually witnessing what I had been told by my college professor.

It got so bad once, that girls kept calling his office line and hanging up. His assistant was not impressed. He was a nice enough guy and I often enjoyed hanging out with him. In my opinion, he was a bit reckless. For example, he got pulled over for speeding one too many times and his driver's license was suspended for six months. He asked me to handle his work cases that required any travel, and I was happy to do it. I always enjoyed being on the road doing interesting things. Was always better being on the road than in the office.

Speaking of the office, it had its own dynamic. Yes, there were cliques and favoritism. Bravado and glory. Unfortunately, it was not an atmosphere where you could really trust anybody. I was not into any of that, so the more time I spent out of the office – the better.

The One We Didn't Figure Out

It was early in the office one morning and I got a report that a corporate aircraft with two people on board had crashed the day before and was recently found. There wasn't much left of the aircraft. It was a nice multimillion-dollar plane with eight seats. No distress signal had been made to air traffic control. The occupants were a male and female. The aircraft seemed to have just fallen from 13,000 feet straight down for no reason. A nearby witness saw the aircraft spiraling towards the ground with the engine roaring. The nose of the airplane was almost straight down.

As you may recall, as accident investigators we look at man or woman, machine, and environment. The crash occurred during climb to cruise altitude after takeoff. The aircraft was halfway to cruise altitude when it started coming straight down. It impacted the ground so hard and so fast that there was virtually nothing left of the aircraft or the occupants.

Needless to say, there wasn't much to go on with this one. But we still had an important job to do. Since there wasn't much left of the airplane and we knew the weather (some light icing reported, but nothing this plane couldn't handle) might only be a small factor, we started our focus on the pilot and passenger.

The pilot was a local lawyer, but also moonlighted as a charter pilot. While he had a good amount of flight time, he was still considered a bit junior in the charter pilot world. This aircraft only required one pilot. The passenger, we later found out, was his girlfriend. An interesting twist ensued as we learned she was living with and engaged to another guy, a dentist. A friend of hers later said the pilot was quoted as saying he could never live with or without her. We engaged a forensic investigator to do a psychological profile on what we knew of both occupants. Neither seemed to suggest suicide. While unfaithful, neither seemed unhappy in any way. No suicide notes were found or alike. The dentist had no idea she was cheating on him.

Since we couldn't pin down any psychological issues with the aircraft occupants, nothing was left of the aircraft itself, witnesses heard the engine was screaming and working well, and weather was not too bad, we were stuck. Maybe a decompression where the aircraft couldn't stay pressurized. Maybe an inflight fire. But no distress call. That only left the possibility that the pilot simply lost control of the airplane while flying in the clouds during climb out.

We later learned that attorneys got involved in the case and a large confidential settlement was eventually reached for the passenger's estate. We heard that the night before the settlement one of the defense attorneys got so drunk at the strategy dinner, he was two hours late to the settlement conference that next morning. A couple months after that, as fate would have it, the plaintiff's attorney was arrested on drug charges. He later pled it out and was sentenced to rehab. Without a full conviction, he was still able to keep his law license. So many strange twists in this case!

Sully

It was a very cold morning, and I was at my office in lower Manhattan. Once again, there were tons of sirens. I got a call from a friend of mine in Chelsea that a plane had crashed into the Hudson. At that point, I had investigated several accidents in the Hudson, most of which were small aircraft. One of them was a mid-air collision between a four-seater single-engine plane and a helicopter that did tours.

This was different. We later learned this was an airliner. An Airbus A320 operated by US Airways had in fact crashed into the Hudson River. Once again, it was bedlam and chaos in Manhattan.

While I was not one of the on-scene investigators, this one obviously hit close to home for many reasons, and I closely monitored the investigation. There is an activation switch on this aircraft that will prevent it from sinking after a water landing. It was activated and the NYPD tied ropes to the plane to prevent it from traveling further away or sinking.

We thankfully learned that all the aircraft occupants made it out relatively safely. The passengers toward the front of the aircraft considered a smooth landing. Passengers in the rear of the aircraft felt it was a very hard landing. The aircraft occupant furthest in the rear was a flight attendant sitting in her jump seat. Part of the floorboard broke and popped up into her leg, which later required surgery. No other significant injuries were reported.

Now comes the investigation. What happened and why? The pilot reported to air traffic control that it had a massive bird strike and lost thrust/power in both engines. The closest airport that was not directly behind them was Teterboro. This is a major airport that supports most of the corporate air traffic going in and out of NYC. Unfortunately, the aircraft glide ratio did not permit enough distance for it to make it to Teterboro. There just was not enough airspeed or altitude.

One theory later brought against the pilots was why did they not turn back immediately toward LaGuardia. As pilots, we are generally trained not to turn the aircraft completely backward after a total loss of engine power. Many times, you will lose too much airspeed and altitude in the violent turn and not make it back to the airport. Also, since this had almost never happened previously, there was some time that went by to realize the dire situation and then start planning accordingly.

Later on, Captain Sully was finally considered a hero. I have had the privilege of spending considerable time with his copilot and the injured flight attendant since the accident. The flight attendant was from Pittsburgh. As you may recall, they were playing in the Super Bowl that year. The CEO visited her in the hospital after surgery and asked if there was anything he could do. She requested that she and the entire crew attend the Super Bowl. Pittsburgh Steelers fans are some of the most dedicated in the entire NFL. She just wanted to go to the game after this tragic event. Well, the CEO organized them to walk onto the field and be honored by all.

It was later learned that the New York airport authority did "a hit" on all the Canadian geese and gassed them to prevent this from ever happening again. Strange, but OK. I understand this could have been much more tragic. As a seasoned air crash investigator, I am very grateful this story ended this way. Airline flying is incredibly safe. The pilots go through the most rigorous training in the world. This was truly a testament to this.

Murder for Hire

It was a gorgeous weekend in upstate New York, and I had the distinct privilege of spending it camping with two special lady friends, a blonde and a brunette. We went on hikes every day and drank by the fire pit every night. There was also a nearby lake with a beach where we would go swimming and boating. The blonde was a tech analyst. The brunette and I went to pilot school together and she eventually got into the astronaut program. She is currently one of the top 100 individuals selected to go to Mars. It really was a spectacular weekend.

On the last day of our outing, we spent it at the beach by the lake. Around 10 a.m. I received a call that there was a multi-fatal plane crash in upstate New York. This was several hours north of us, and I determined the best course of action would be to drop the girls off as quick as possible, hop in the airplane and fly up there to determine what happened.

As I collected more information about the plane crash, I learned that the plane was with an outfit I had worked with before as they previously had a minor landing event. I knew the folks there well and I liked them a lot. I was also really surprised that a corporate aircraft on a charter flight crashed on such a beautiful day. Especially with this organization that operated the airplane that crashed.

After dropping the girls off and hopping in the "company plane," I made the trek upstate with no problems. When I landed, that is where the problems began. There was nobody at the airport. It was just a runway and a grassy area for parking. No other structures whatsoever. I also had a terrible cellphone signal. So I decided to start walking into town until somebody finally picked me up and drove me to the only hotel in town. There was no rental car place, so I called a local sheriff's office for advice. Thankfully, one cop in a two-cop town was gracious enough to take me to the crash site.

He was a unique police officer, young, and seemed oftentimes bored. I had always wanted to be a police officer, so I was very interested in his job. He even let me play with his lights and sirens. He later admitted to me that he was an alcoholic and there wasn't much else to do in that town.

The plane crashed into a mountain near the airport. We went right to the crash site after a medium-distance hike, we didn't find much. There was very little left of the airplane nor the occupants. Nothing bigger than the size of a basketball. All that could be done on scene was document what very little was left. Daylight faded fast.

The next day I needed to get to the crash site and spend some real time there. I knew I would not be able to utilize the local police department again for that amount of time. So, I had to get creative. There was a Toyota car dealership in town. I told them I was interested in a Toyota minivan and wanted to rent it for the day to try it out. They literally just loaned it to me and that's how I got to the crash site.

Again, I did not find much and the bugs were out like crazy. You could literally put your hand down your shirt and pull out a bunch of flies. Other investigators on scene that day did find a gun clip with two rounds expired. No gun, just a gun clip. Weird.

I eventually made it back to the plane and back to the office the next day. I also later found out it was the owner of the company that was flying the airplane. My buddy Mason, whom I had previously enjoyed working with and respected, was the pilot. However, who was the single passenger? The airplane required two pilots.

The accident happened on a Sunday. By Thursday I received a call from the FBI. As you know by now, I have worked with them in the past and had a good rapport with them. They shared with me that the passenger was wanted on fraud charges and recently took out a large life insurance policy. I also learned from the charter company that the passenger held himself out to be a pilot, although he was not.

While the exact cause of the accident was never completely determined, it was believed the passenger shot the pilot and himself before aiming the airplane toward a mountain. Mason's daughter worked for her father at the charter company. I called her to give her some of the facts and she said to me, "Don't tell me my father was murdered." I told her we could not confirm that, but she should review the case further with the FBI.

Mason was a good guy. Trusting and would do anything for anyone. It's a shame this happened. He will always be missed.

Planes and Connecticut Don't Mix

Many times, I find wonderful vacation destinations from doing an accident investigation there. A small town in Connecticut ended up becoming one of them. It's a sleepy coastal town on the southeastern shore of Connecticut on the Long Island Sound. It is near Mystic, Connecticut, and yes, there really is a Mystic Pizza there. Just like the movie! It's also a major submarine manufacturing facility.

It has an airport with a short runway for jets, around 5,000 feet long. While that may seem long to most people, Learjets typically require 6,000 feet or more and good conditions. Unfortunately, I've had to investigate two jet accidents and one prop/piston-powered airplane accident in the area. All of which were accidents while on landing approach.

The first jet accident was a reposition trip from Long Island. A reposition flight means the jet is flying empty to go pick up passengers. While this jet was coming in for landing it seemed to be too high and too fast to make a proper landing and the pilots decided to go around and come back to land again. This is proper protocol given the situation. However, the pilots left the gear down and brought the flaps up while reducing power. They also put the aircraft into a 65-degree bank turn. This caused an aerodynamic stall (called an accelerated stall) and the aircraft sank into two houses, setting one of the homes on fire. The aircraft then struck a river, and the pilots were ejected from the aircraft due to the brunt deceleration force. An older lady in one of the houses that was on fire had to jump out of the window to survive. Thankfully, she only had minor injuries. The only fatalities were the two pilots.

Now came the investigation. This particular aircraft was equipped with a cockpit voice recorder and flight data recorder (CVR and FDR). We quickly learned the gear was never raised which increased drag on the airplane significantly, and the flaps were raised – which dramatically increases the stall speed potential. As a plane is in a turn, stall speed usually increases as well. Thus, the aircraft entered into an accelerated stall. At this low of an altitude, it was not recoverable. What the pilots should have done is climb back into the clouds, then stabilize the airplane, then execute the approach properly back to the airport. While they were flying this plane like a fighter, they were heard on the cockpit voice recorder yelling "Yahoo!" Unfortunately their showboating forced them to pay the ultimate price.

About a year later a Cessna with four occupants was shooting the approach for landing and crashed into the water about a mile short of the runway. It was later learned they were foreign pilots who were not familiar with the area. They got low and slow on the approach and crashed into the water. The day of the accident, the Coast Guard tied a balloon to a cinder block to mark where the plane was. A couple days later when we got there with a barge and a crane to lift the airplane, it was gone. The tides and currents were so strong they moved the airplane to a point where we could no longer find it. It took us a week to finally locate it again so we could retrieve it. Back to the accident investigation. The lesson here is if a pilot is unfamiliar with a local airport environment, get training on it first. This especially holds true for tricky airports like this one.

Another crash was a flight instructor and a student pilot attempting to land there at night. They simply came in too low and struck a tree. However, they survived. But why? Why did they survive? The interesting factor is that their airplane had airbags equipped in it. This saved their lives. They were sprayed with fuel that leaked from the damaged wings and stayed in the cockpit nearly overnight. Other than mild hyperthermia, they were found the next morning in good order. If you have recently been on an airliner and saw baggy seatbelts in the first row, those are actual airbags!

The next accident I investigated there was a jet from Atlantic City. It was a chartered casino trip with two passengers gambling in Atlantic City, then going to casinos in Connecticut. The aircraft was coming down on approach in mist and fog but landed short of the runway and struck a stanchion light. The jet cartwheeled in the water and came to rest upside down. Both pilots were killed; however, both passengers were able to successfully exit the aircraft with minor injuries and ended up at the casinos that night after limousines picked them up from the hospital. I guess they were feeling lucky.

The night before the pilot told his wife that the runway was short and would be wet from rain and fog. This most likely meant he would come in low during the approach to try to land on the very first part of the runway that he could. Most jets land about 15 percent down the runway at the touchdown markers. So, we surmised through our investigation the pilot came in low and got way too low before striking the approach lights. When the aircraft cartwheeled in the water, the wing broke off and went through the cockpit striking both pilots in the head, killing them both.

Once again, we had to spend a couple days on a barge with a crane to recover the aircraft. There was also a lot of currency on top of the water that we recovered as well for the police to get back to the passengers. Looking back, if the aircraft traditionally did not do well on short runways that were contaminated with rainwater, they should've gone to a larger airport such as Providence, Rhode Island.

When pilots push their aircraft and personal limits, this is what can happen. Unfortunately, it happened three times during my career in CT. To this day I love visiting the area. It's a very peaceful and tranquil waterfront coastal town in New England.

Ever Play Baseball in Manhattan?

One of the interesting things about Manhattan is everything you see. I remember the first trip I did alone to Manhattan was back when I was with American Airlines. We landed at LaGuardia late at night and made our way to our hotel in midtown. About two blocks from the hotel near Penn Station was a person being zipped up in a body bag. It was midnight. I said to myself, "Yup, I'm in NYC."

Now for the baseball story. At this point, I had experienced the terrorist attack of September 11, 2001, the blackout of August 14, 2003, Captain Sully on the Hudson, etc. In October 2006, while at my desk in Manhattan, I got word that a plane crashed into a building in the Upper East Side near LaGuardia airspace.

Weather that day was not too bad. The aircraft was a single-engine four-seater with two people on board. So, I hopped on the subway and made good time getting on scene. I went to the NYPD command center and started working with the other investigators who were arriving on scene.

We quickly learned a pilot and a flight instructor were on board the airplane. The aircraft was last seen heading north above the East River and made a U-turn to the left and struck a building around the 40th floor. The aircraft went into a penthouse apartment – then fell to the street below.

The crazy thing is why did this happen? Was this intentional? A couple years before, someone in Texas loaded up an airplane with kerosene and flew it into the local IRS building. Another flew his airplane into a downtown Tampa building with a suicide note in his pocket. Neither caused considerable damage, but were both noteworthy. So what happened here?

The news quickly picked up that the pilot was a pitcher for the Majors. He was a young pitcher and fairly new to the league but had a promising career. The flight instructor he was with was from California and fairly experienced.

The aircraft was a low wing airplane, meaning the wing was below you during flight versus the wing being above your seat. No problems were reported to air traffic control, and nothing wrong was found with the airplane.

Things became more obvious when we looked at the radar track and an aviation map. As the aircraft was heading north above the East River, it was fast approaching LaGuardia's airspace. Hence the U-turn. The plane did not turn sharp enough, and the low wing blocked the view of the nearby building. They likely never knew the building could have ended up just in front of the wing during the tight turn.

There is an interesting concept term that may be applicable here. It is called boxed canyon theory. In other words, imagine if you were flying through a canyon wherein the walls of the canyon are closing in all around you. You try to make a U-turn but it's too late as the walls are too tight around you. Almost like flying into the corner of a triangle. Similar concept here. Your clearance along the East River becomes very tight as you head north, and LaGuardia's busy airspace is right there.

Thankfully the owners of the penthouse apartment were not home when the crash occurred. I later found out it was a couple that lived there, both doctors. The wife had been in the news before as she was at the Macy's Thanksgiving Day Parade the year before when one of the floats struck a post, which fell on top of her – causing injury. She was a lucky person. Reminded me of a cat with 9 lives!

Divorce an Option

The summer of 2001 was beautiful in the Northeast. Lots of sun-filled blue skies with white puffy clouds in partial view. Very little rain and fairly mild summer temperatures. One Saturday morning in late August I received notification of a plane crashing into a house. This immediately sounded strange as weather was really good. But as you may recall, accident investigators look at every aspect of any crash.

With our suspicions raised, we further looked into this crash before responding. The news showed some photos and there was not much left of the aircraft or the house it crashed into. We therefore elected not to respond until more was known.

Sure enough, by that evening, the news was reporting that the pilot crashed into his own home the morning after he was served with a restraining order from his wife. So yes, this was an intentional act of an attempted murder/suicide.

We later learned the wife of the pilot had been in an abusive relationship for decades. The pilot owned a successful business, which was rumored that the only reason why he owned a business is because he did not work well with others. He was a perfectionist and awfully critical of others constantly. Everything always had to be his way.

The pilot's wife had suspicions and concerns that a divorce could make the pilot violent. So, the night the pilot was served with a restraining order, his wife packed up herself and their daughter, then moved into a hotel. That decision likely saved their lives.

The 9/11/2001 terror attack of the World Trade Center in New York happened a couple weeks after this, and the wife felt responsible in some way since it was uncommon for an aircraft to fly into a building until then. At least not intentionally. This likely was a factor of PTSD, which she has mostly overcome. These days she wants people, both men and women, to know that there is help if you are in a bad relationship and need to get assistance.

While this pilot likely thought he was killing his family along with himself, he was wrong. The wife of the pilot was later quoted that toward the end of her marriage she reached out for professional help and needed validation of her feelings of concern before she could take action. She wishes she had reached out to someone sooner, and to this day she volunteers for domestic violence assistance. She eventually remarried and they adopted another child. Her daughter completed college and got married.

If you ever consider harming yourself and/or others, please seek help immediately. You may feel that ending your life or someone else's life may solve all your problems. The only thing that does is eliminate the chance to solve them.

Snow Globe in a Hangar

2010 was a very cold winter with tons of snow. And it wasn't just the white, fluffy snow; no, it was the heavy, thick icy snow almost all winter. I usually do very well on snow and ice, including skiing. This particular winter, I actually fell twice from slipping on ice. It was everywhere.

Well, come to find out I was not going to be the only victim of this horrible winter. Yes, we all know the white, fluffy snow that everyone loves and plays with in the movies. What many don't realize that dark, thick snow is very heavy. There have been several instances where aircraft get too much snow and ice on its tail during a storm, and the tail will strike the ground. This is caused by a large amount of new weight on the tail shifting the center of gravity aft/rearward to a point where the aircraft nose comes off the ground and the tail strikes the ground while the aircraft is parked.

One winter storm came through the second week of February in 2010. It was a very damp, icy, and snowy storm with blustery winds. A true nor'easter. From southern Virginia through northern Maine, nothing was spared. Everything was covered in snow and ice, the kind of ice it takes 20 minutes to chip off the window of your car.

Washington, D.C., received the brunt of the storm. A hangar with many jet aircraft ended up with the roof caving in from the weight of the snow. A hangar beam failed and pushed down onto the tails of the aircraft parked inside of the hangar. This caused several jets to be in a nose-up position, as if they were taking off. There were many aircraft that were damaged from the hangar collapsing due to the weight of the snow.

The heavy snow in the Washington, D.C., area resulted in major damage to the roof of the Dulles hangar. The hangar roof collapsed under the weight of as much as three feet of snow. Though five people were inside at the time, no one was injured. Two Bombardier Global Expresses and a Gulfstream G550 were seriously damaged when the roof collapsed. Damage to the building was described as "significant." Besides the destruction at Dulles, snow also caused damage at nearby Manassas Airport in Virginia.

Most of the aircraft were owned by major companies and were not economically repairable. It was later determined the hangar was not constructed properly and the manufacturer was no longer in business when the collapse occurred.

Drugs For Everyone

One of the things I love most about this job is you never know what is going to happen, nor when. It was a normal day in the office. Catching up on some paperwork. Making some phone calls. It was late spring, and the type of weather which usually happens that time of year is partly cloudy with the occasional afternoon storms. Sure enough, we received notification of an aircraft that broke up in flight over Indiana. Two fatalities were reported (pilot and passenger, who was also a pilot). The aircraft was mostly destroyed.

Typically, when an aircraft experiences an in-flight breakup (wings or tail often departs the aircraft), it is usually attributed to one of two factors. The first, over-stressing the aircraft during flight. For example, if the pilot pulls too hard on the controls while in a stressed airframe situation such as a storm or tight turn, the airframe can become over-stressed to the point where it starts to break up in flight. The second is mechanical breakdown. This can occur from corrosion, improper maintenance, improper parts, lack of maintenance, etc. Luckily, both of these are extraordinarily rare.

As investigators, we are trained not to only focus on one probable cause or contributing factor, as it is important to rule out all possible causes first that are unrelated. For example, if we find a bunch of fuel at the crash site, the plane likely did not completely run out of fuel.

In this particular investigation, it is easy to assume the pilot either overstressed the aircraft or the wing failed due to mechanical reasons. However, please remember it is important to consider any possible factor in any crash.

As we continued the investigation, we later found out that there were some witnesses who saw the aircraft going in and out of the clouds at odd angles. This was strange, considering the pilot did not hold a commercial pilot certificate with an instrument rating. An instrument rating allows a pilot to fly into the clouds and navigate via instruments only. Without one, a pilot is prohibited from flying into the clouds. So why was this aircraft viewed flying in and out of the clouds at strange angles?

As part of most aircraft accident investigations, there is the on-scene inspection of the aircraft and a later wreckage inspection at a location where the aircraft is moved for final storage, usually within a couple months of the crash. The post-crash wreckage inspection is usually much more detailed and has a few more parties involved, such as attorneys and experts. This certainly was the case in this instance, as the passenger's estate/family was now suing the last maintenance facility that worked on the aircraft – pointing toward this being a maintenance issue versus pilot error.

Since investigators typically have more time to spend at the wreckage inspection than the on-scene crash site, investigators really try hard to not leave any stone unturned. During the wreckage inspection at the storage facility, there was no strong evidence found to suggest there was any corrosion of the wing at or near the failure point or mechanical breakdown.

So that still leaves us with a question as to why the pilots decided to fly the aircraft in and out of the clouds in an odd manner when they were not licensed to do so. The investigation continued. Sometimes, there are electronics on board which can further assist the investigation, such as camcorders, cellphones, GPS devices, etc. We decided to check the luggage for same.

We were not able to locate any additional electronics in the pilot's personal luggage. However, we did find lots of drugs. Cocaine, illegal pills, etc. You name it. Around this time the toxicology reports came back and both pilots were positive for illicit narcotics. And it now made sense that witnesses on the ground viewed the aircraft flying erratically in and out of the clouds before it crashed. It appears the pilots overstressed the aircraft, which caused the airframe to fail. Poor decision-making. Case closed.

Big Plane for a Student Pilot

For many of us pilots, we can look at almost any airplane in the sky and tell you exactly what it is. We of course understand that to most others, a plane is a plane. Some may think that any training airplane is a Cessna. That is funny to us pilots, as we know that Cessna made the fastest jet in the world at one time, the Citation X (ten), as well as many other popular jets. Some people even know the difference between a Piper, Cessna, and Cirrus. The point to this is it takes a lot of training to be able to fly different types of aircraft proficiently.

I received a call one afternoon that a pilot flipped over his airplane on landing. He came in a little too fast on a dirt runway and hit the brakes too hard. The nose wheel dug into the ground and the aircraft flipped over itself. The pilot was uninjured, and naturally the aircraft was significantly damaged.

Remember, during any accident investigation we look at man or woman, machine, and environment. We try to determine what happened, why it happened, and how to prevent it from happening in the future. The weather that day was pretty good. The dirt strip may have been a little muddy from recent rain, and that could have played a factor into why the nose wheel collapsed after digging into the dirt.

The aircraft is a Cessna 210. While it is a single-engine propeller-driven Cessna, it is very powerful (turbocharged) and complex. Overall, it has a very good track record and is reliable.

So, after investigating the machine and the environment, it's time to review the pilot info. This can be as simple as documenting the pilot's current qualifications, to reviewing all the pilot flight history. Usually, this information is gained when the pilot fills out an accident report. There is an entire page for the pilot qualifications and flying history.

In this accident I decided to verbally ask the pilot some basic questions about his qualifications. He had several hundred hours of flying time. When I asked him about his license, he advised he was a student pilot. This was very unexpected considering the caliber of the aircraft he was flying and that he had several hundred hours of flight time. He claimed he just never got around to getting his actual pilot's license. After further investigation, we further found out he lied on his insurance application as well.

To this day I am still dumbfounded as to why someone would purchase and fly a sophisticated aircraft without first attaining the proper licensing. I guess that's why it ended up in this book...

Helicopter Life

In general, I am not a helicopter guy, not a helicopter pilot, and try to avoid them for the most part. I also freely admit therefore I am not the world's best helicopter accident investigator. Unfortunately, I have had my fair share of investigations involving helicopters. From mid-air collisions to inflight break-ups, to collisions with terrain. There are three helicopter accidents that stand out to me.

I had recently been transferred from New York to Atlanta. Thus, when a helicopter went down in Louisiana, it was assigned to me to investigate. There were two people on board and no survivors. The helicopter crashed and burned as well. It was in a very remote location, on a small island in a bayou. There were two ways to get there, either via a one-day boat trip or a one-hour helicopter ride. While I had some choice in the matter, I chose the helicopter route. I had only been in a helicopter a couple short times before this. While I was a little nervous, I was also excited to see the southern Louisiana landscape. It was eerie, however, flying the exact route in the exact helicopter type the accident pilots recently flew to their demise.

Thankfully, I was with an older experienced pilot and the flight there was uneventful. The two pilots on board the accident helicopter hit so hard they were ejected out of their seatbelts. A couple hours into the investigation, a thick fog temporarily rolled in. The fog came in around the same time the accident happened the day before. I wonder if this is what the accident pilots experienced.

While the helicopter was in pieces and burned, we could not find anything wrong with it. We later determined the pilots started flying lower and lower to avoid the fog before eventually crashing into a tree during a turn.

It was a sobering experience flying the exact trip the accident pilot flew just the day before and seeing the weather roll in that likely caused them to crash. I later learned the pilot that flew me to and from the site complained that I asked too many questions. I was trying to find out more about helicopter life, but I understand it was hard on him losing a friend. If you are in this business long enough, it likely will happen.

The next fatal helicopter investigation I did was an experienced pilot for a major company taking VIP guests from a party nearby for a ride. After the guests had all gone up for a ride, a new hire was offered the opportunity and took it. She was a 19-year-old bartender. She had just started college and took this job to make some extra cash.

Witnesses saw the helicopter flying just above the tree line at high speeds. When it came near powerlines, it dipped down to try to miss them, but was unable. The helicopter struck the powerlines and then a tree before falling to the ground. Both occupants survived the initial crash but died soon thereafter due to head trauma.

So here was this experienced pilot who took a 19-year-old girl on her first helicopter ride and was hotdogging just above the tree line to show off. I have never flown an aircraft low and slow, and never will. Altitude is often your friend when flying.

The last helicopter crash that comes to mind is one over the bay near Tampa, Florida. Witnesses at a nearby beach saw the helicopter flying normally before suddenly dropping off straight down into the water. The pilot was the sole occupant on board and did not survive. The aircraft ended up in about 20 feet of water and was swept away by multiple currents. Part of the helicopter was found, except for the two main rotor blades.

The pilot was very experienced and also flew for a major airline. The helicopter had a very good maintenance record and had recently had an inspection. This accident was baffling in so much the helicopter simply flew straight down out of the sky. Did the pilot have a stroke or a heart attack? The autopsy revealed no health issues with the pilot.

Maybe the helicopter struck a bird? We really wanted to figure out what happened. Sure enough, somebody was filming nearby and caught the accident. They called the local police to tell them, who advised us that this was available. Albeit far away, we had video of the accident! We dissected this video frame by frame. In one of the frames the helicopter appeared normal. In the very next frame, the rotor blades were breaking off and the helicopter was heading straight down.

But what would cause the rotor blades to break off during flight? Again, was it a bird strike? Bad maintenance? It just passed a major routine inspection. This really made no sense. We needed to find the two rotor blades and we were so far unable.

We had run out of time and resources. The recovery investigation was now over. However, this did not sit well with me. My sister lived nearby and was married to a diver. I had offered to pay him to look for the rotor blades, but his schedule with his new job did not permit it. So, we got creative and offered a reward for any diver that found them. Within a week, they were located!

What we then discovered made the case take an incredible twist. Part of the rotor blades were missing. It was like somebody cut out a square of each rotor blade. This truly did not make sense. After reviewing this with the factory, the rotor blade is made in several pieces. Part of the blade departed the rest of the blade; this was due to a phenomenon called delamination. It can happen with the pilot over stresses the aircraft resulting in a state called mast bumping. The blades essentially no longer spin freely and start rubbing against the center mast attach points. They eventually fail in-flight. While this is rare, their design eventually was improved to ensure this never happens again.

I just realized I completed this chapter on the one-year anniversary of the Kobe Bryant crash. All accidents are unfortunate. However, some of them result in safer practices that save lives in the future.

To Present or Not to Present, That Is the Question

Around my second or third year in the business, I was asked if I wanted to join the federal safety team as an accident investigation presenter. I could start with a simple topic about aviation, such as wake turbulence. This is a phenomenon wherein an aircraft disrupts the air so much that the air behind it in trail is disturbed to the point it could be dangerous to other aircraft. It's kind of like the wake a boat leaves as it goes by. Usually not a concern, but can be very dangerous at times. They are often referred to as lenticular tornadoes. So, pilots are trained, and aircraft are directed to avoid this dangerous situation.

While this seems like an easy topic for me to tackle in front of an audience, it just wasn't for me at the time. I was still adjusting into my new role as an accident investigator, very busy, and had little free time. Perhaps some maturity needed to develop a little bit as well before I was ready. I was also working part-time as an airport passenger transportation driver to try to make ends meet. New York was very expensive, and I was not making a lot of money at the time.

A couple of years later I was approached again about joining the volunteer federal safety team as a presenter. At this point in my career, I had seen dozens of accidents that I felt could have been preventable. So now I firmly had a message I wanted to deliver to the aviation community.

With the help of some of my colleagues, I developed a presentation of five or six actual accidents to present to an audience of pilots. I would first show the audience how we investigate plane crashes, then make them be the investigators while I fed them facts about each accident. The audience would learn how to determine what happened, why it happened, and how to prevent it from happening in the future. It was my hope by doing so that the audience would learn from other's mistakes and not get themselves into the same position.

I took the show on the road, and it was a hit! It was not long before word got around and most of my presentations were standing room only. Before I knew it, I was being invited to present out of state and eventually around the entire Northeast. As a pilot, many of these presentations I could simply fly to and fly back from myself.

It seems like I inspired a couple audience members to become accident investigators themselves. This brought me great joy and I was happy to help. Most of them did not stick with it as this job can be taxing, both with a challenging schedule and emotionally hard. However, I found that what I was doing was rewarding. I also hope that I helped save a life or two along the way as well. Within a few short years, I gave dozens of presentations to hundreds of pilots.

Apparently, my efforts in the industry did not go unnoticed. I was nominated for a national award recognition for my volunteer service. Within a couple months of my nomination, I was awarded the eastern region volunteer safety representative of the year. This was an honor and I got to attend an award ceremony for it. This inspired me to do even more.

One of the largest airshows in the country is the Experimental Aircraft Association Fly-In located in Oshkosh, Wisconsin. I was lucky enough to go there in the year 2000 and loved it. 10,000 planes and 600,000 people. The presenters there were amazing, the best of the best. It would be ten years until I went back there.

In 2011, I was invited to be a presenter at Oshkosh. Man was I honored! By far the majority of the general aviation pilot community goes to Oshkosh. This would be my opportunity to share the lessons I've learned with others to make aviation even safer. At this point I had performed dozens of presentations for hundreds of pilots.

I was assigned a date, time, and large pavilion with full video and audio capabilities. As pilots and alike trickled in, I nearly had a full house before I even started. I gave the same presentation format I had done in the past and it was a big hit. Little did I know, but this led to me being a presenter at Oshkosh for almost 10 years in a row. Eventually, I developed a following and would see familiar faces in my audience year after year. That was very humbling to me.

One of my audience members was a fellow accident investigator. He must have been impressed, as he became my boss. That's right, these presentations helped lead me to where I am today. I even got invited to the VIP tent during the military and civilian airshow performers, catered celebrity dinners, and a few other amazing perks. But in the end, what is most important to me is that my audience enjoyed it and learned something.

Animals and Airplanes Don't Mix

As mentioned in Chapter 8, birds can be a real problem for airplanes. Not just in flight, but also on the ground. They make nests in planes, particularly in and around the engines, as they are often warm – even hours after a flight. Thankfully in my 30 years of flying, I have never had a bird strike. But man have I come close.

There was one particular bird strike of interest I investigated which I will always remember. It was a single pilot flying a medium-sized general aviation aircraft who had a turkey vulture come through the windshield while he was flying. It immediately shattered the window and slapped the pilot hard in the face. The pilot was covered in blood and guts, including in his eyes. The quiet cockpit immediately turned into a wind tunnel of hell. The pilot wiped the guts and blood from his eyes with his shirt and struggled to maintain control of the airplane. His headset was knocked off his head so hard it was unusable. Thankfully, he was able to point his plane toward an airport and make it onto the runway. It was like a scene out of a movie. But the pilot remained calm and knew what to do.

One time when I was flying in New York, a couple hornets had gotten into the airplane and started buzzing around and stinging everything. I was alone, thank God. So now I must figure out what to do. I was about 10 miles away from landing at my home base airport. So I decided to tell air traffic control that I would be off the air for a minute. The aircraft I was flying was not pressurized, so I opened the window and sure enough the pressure differential was enough to suck the hornets out of the airplane. Phew! At least it was better than a guy in Australia who had a snake crawl out from underneath the panel. He grabbed it and landed with the snake in his hand until he could throw it out the window.

Sometimes we have an aircraft strike a deer on takeoff or landing. This instance was a bit different. An aircraft had landed, and a deer ran smack into the side of it. It may have been chasing another deer and simply didn't see the plane. The deer spun around and scampered back into the woods. The aircraft was damaged, but thankfully not significantly.

Often when pilots travel, they bring drinks and snacks in the plane. This is a very common practice. Naturally, if pilots don't finish all they brought, many times they leave it in the airplane for the next trip. Well, it's not just pilots that are attracted to that. Yes, I am referring to bears. They love to tear up an airplane in an attempt to get food. This can prove to be very costly. And it's not just bears. In Africa a herd of elephants decided to do the same thing with a jet. $100,000 in repairs later, the plane was good to go again.

Within the last year, I have personally seen deer, coyote, fox, birds, and turtles on airport property. Airports typically are large fields in somewhat rural areas, and animals think it is their oasis.

The Barefoot Bandit

One of the things I love about my job is promoting aviation safety, as previously mentioned. There are many groups and organizations that are dedicated to doing just this. In my opinion, it really paid off as accident rates went down considerably over a ten-year period (2000–2010). Also attributable to this success was amazing developments in technology. It used to be in the good old days you call flight service for a weather briefing and launch on your flight looking out for other airplanes and weather the entire time. Now, most airplanes have weather radar capabilities of some sort and basic traffic collision avoidance systems. Really helps pilots make good decisions in the cockpit. This in turn helps reduce accidents and save lives.

So, my job thankfully was a little bit slower in 2010 than when I started in the 1990s. Thus, when a report of a stolen aircraft hit my desk, I was ready to take it on with any resource possible. An aircraft was stolen in Washington State, and with radar technology, we were able to track it eastbound. We were a day or two behind it, but not by much. Interestingly, every airport he landed at; the thief would take another airplane. Even more interesting, we learned he was barefoot and would break into local homes and businesses taking mostly food and supplies. He wasn't the best pilot as some of the aircraft ended up damaged.

He kept flying eastbound until he eventually hit Florida. He had stolen several aircraft at this point. He did not have means for purchasing fuel, so he just kept stealing airplanes that had fuel in them. Once in Florida, authorities were hot on his tail.

So he probably thought if he leaves the country, he cannot be extradited back to the United States. And so, he stole another aircraft and flew it to the Bahamas. By this point he had become well known and the news media picked up the story. He was coined "the barefoot bandit" for obvious reasons. He was also posting his journey on social media and actually had followers supporting him! He was 17. And NOT a pilot!!!

Naturally, Bahamian law enforcement was on high alert for this kid.

So when he went to the island of Grand Bahama, he stole a boat and eventually was caught by a security guard in a marina. And yes, he was extradited back to the United States.

We later found out he learned how to fly using Microsoft Flight Simulator. Never had an actual flight lesson. That's it. Smart kid, but went about it the wrong way. We later learned he came from a broken home. He ended up having to pay restitution and serve many years of probation. Last I heard he was working for the lawyer that defended him and using that money to pay back his debt. He later started a crowdfunding campaign on gofundme.com to pay for his flight lessons to become a professional pilot. That quickly was shut down…

My Passenger Fell Out

It was November 2013. I heard about a story of a passenger falling out of an airplane. It quickly became a well-known story, as the pilot's transmission over air traffic control was quite unique. He said, "My passenger fell out of the airplane." Whattt?

The air traffic control tower responded back to the pilot and asked if a passenger fell out of the airplane. The pilot responded by saying that was correct. An emergency was declared as the pilot now had to land the airplane with the door open.

The aircraft was approximately six miles away from the airport at 1,800 feet. This is a very typical altitude for this distance from the airport. A fall from this altitude, even into the water, is not survivable.

What happened here? The pilot made very clear and calm transmissions to the tower. The passenger's body was eventually found. He was 42 years old and pronounced dead upon discovery.

Homicide detectives worked the case. The pilot was very cooperative. The aircraft sustained minor damage from the event. Eventually, the matter was considered a suicide. This was not the first time this had happened. A charter passenger jumped from an airplane about 20 years prior. Same area too. This was definitely one of the more interesting cases I investigated. The pilot later told me he simply was taking a friend/neighbor for a ride in his airplane. Little did he know what would happen.

Ice and Airplanes

If you have ever flown on an airliner in the wintertime, you likely may have noticed the airplane is deiced while you're aboard or just before boarding. A deicing agent is sprayed onto the wings, fuselage, and near the engines. Well, why is this performed?

Let's take a paper airplane. If you throw the paper airplane through the air, it will glide for some time. However, if you add particles (i.e. drag) to the paper airplane it will not glide as far. In other words, any disruption of airflow over the wings of the aircraft can greatly reduce the ability of the wing to produce lift. Therefore, it is imperative to have little to no ice or frost on the airplane prior to takeoff.

Once an aircraft is in the air, many of them (especially airliners) have heated wings and tails. That way if icing conditions are encountered, the aircraft can maintain flight and prevent the ice from accumulating on it. Many smaller general aviation aircraft do not have this or have a very limited anti-ice function. That is why they drill into you during flight training to avoid icing conditions as much as possible.

Unfortunately, some pilots knowingly or unknowingly fly into icing conditions. This can be particularly dangerous in precipitation-rich storms that are near freezing temperature. A single storm like this can take out several aircraft. What happens during flight in these conditions is the aircraft slowly starts accumulating ice until it has accumulated so much it can no longer be controlled by the pilot. This was the case late one fall in the Bible Belt/Southern states.

The first one started in Tennessee. It was a cold day with very dark, gray clouds hanging low. A small twin-engine airplane with only the pilot on board was trying to land at a municipal airport near Nashville. The aircraft attempted an approach in conditions where weather was deteriorating with misty rain and even cooler temperatures. The aircraft aborted the approach and lined up to try again. On the second attempted approach, the aircraft lost control and crashed inverted into trees. The pilot did not survive.

Around the same time in West Texas, a pilot and his son in a medium-sized twin-engine airplane encountered icing at 13,000 feet. The aircraft came almost straight down and crashed into a prairie in a remote area about 50 miles away from Amarillo. I was able to get on scene for this accident the very next morning. We quickly figured out what occurred structurally. Parts of the aircraft were separated over a large area. This is usually indicative of an in-flight breakup, meaning the aircraft came apart during flight. This is usually caused by the pilot's excessive control inputs which causes major stress on the airframe until it fails.

Two more accidents occurred from the same storm, one near Atlanta and the other in New Jersey. Both, ironically, crashed into a highway. Thankfully nobody on the ground was hurt. In total, almost a dozen aircraft would have crashed from that one winter storm with nearly 20 lives lost...

So why does this happen? If pilots know not to fly into icing conditions, why do they? During the normal progression of your flight training as a pilot, you earn your private pilot license first, then usually your instrument license comes next. Just as it sounds, your instrument license (which is actually a rating) allows you to fly in the clouds via use of your instruments for reference of navigation only.

Well, during the winter months that means you could possibly encounter icing conditions (in clouds near freezing conditions). There are different levels of severity of icing from light, moderate, to severe. Before a flight, a pilot is trained to obtain a weather briefing. Part of the weather briefing should include if icing encounters are likely or not, and to what severity. It also should include pilot reports from pilots (called PIREPS) that flew through these conditions recently and what they reported to air traffic control or flight service. As you can imagine, icing condition forecasting can be tricky.

While it's best to try to avoid icing conditions at all costs, some aircraft are equipped for it – which is called certified for flight into known icing. However, it only works if the equipment functions as it should, and the pilot operates it properly. With the advent of in-flight radar capabilities both on the instrument panel and on a pilot's iPad, thankfully icing accidents seem to have greatly reduced over the last few years.

Flight School Training, Anyone?

There are several different ways a person can receive flight training. The most basic is a flight instructor that rents or owns an airplane for instruction, next is a small mom-and-pop local airport flight school, and the last is a national flight school. Each of them has their own unique characteristics, and believe it or not, their own rules/regulations they must follow. Federal aviation regulation (FAR) part 61 is a subset of regulations designated for its smaller and less structured flight training facilities. Federal aviation regulation part 141 is a subset of regulations designated for larger flight schools. 141 (type of training, not a numerical amount per se) schools use a very structured syllabus that lays out nearly every bit of your flight training.

On a Saturday afternoon in April, I received a call that a plane had crashed near my local airport. On board was an instructor and student. No survivors were found. I remember thinking the weather was quite poor that day, as this was the first day in late winter that I heard thunder and saw lightning. That usually signified the beginning transition into spring.

What were a flight instructor and student doing flying in conditions like this? Where exactly did they crash and what went wrong? Was an emergency declared with air traffic control (ATC)? Did they run out of fuel? What condition was the aircraft in? By the time I received the notification, it was too late in the evening to launch on it. So, it was decided to start the investigation the very first thing the next morning.

The weather thankfully was a little better the next day. The aircraft appeared to have hit a tree about a mile away from the airport and then came almost straight down. The occupants likely died of blunt force trauma to the head. During the on-scene investigation, we document the aircraft, as much as we can about air traffic control, maintenance, and perishable evidence, such as ice or snow.

The ground was muddy from all the rain we recently had. Another indication that weather was a likely factor. The aircraft was not as badly damaged as I thought it would be. We were able to document everything in the cockpit, and around the airplane as well. It appeared to have hit the treetops at a steep angle before coming to rest.

We later learned the student pilot was new to flight training. He had not soloed an airplane yet but wanted to be an airline pilot one day. The instructor met all the qualifications to be one but was not as experienced as most. The weather report at the time of the accident showed low ceilings, fog, mist, and some embedded thunderstorms. Basically, the entire area was socked in.

So why did the student and instructor launch in such dismal conditions? Perhaps one may argue that since the student wants to become an airline pilot, this type of experience would be good to have. On the other hand, this is a national flight school and flight training is all they do. There is a sense of a strong directive to keep the airplanes flying as much as possible. The aircraft logbooks showed that it did fly a lot. In general, that means sometimes in bad weather.

At the end of the day, the accident report showed the primary factor was the flight instructor's inability to properly shoot an approach and subsequently he crashed into trees. A contributing factor was the decision to take a primary flight student into instrument conditions/bad weather. ATC even issued a low altitude alert, which was acknowledged by the instructor but with no action taken by the instructor.

Later on, a lawsuit was brought forth by the family against the flight school with allegations that the flight school sometimes operated recklessly to make a profit. It was rumored that a settlement was eventually reached.

Not an Emu, an EMAS

OK, so what in the hell is an EMAS? To be specific, it is an engineered materials arresting system. Since that explains it, I hope you enjoyed this chapter.

Kidding! So, what exactly is this and what exactly does it do? Have you ever heard of an aircraft running off the end of the runway either on takeoff or on landing? If you have, then you speak my language. It's called an overrun. If you haven't heard of this, trust me, it happens way more often than you think. It often results in serious damage, injuries, and unfortunately – fatalities. So, the FAA had to come up with something to help the situation.

What do we do to prevent aircraft from going off the end of the runway? Do we use big nets like an aircraft carrier might use to keep aircraft from going overboard? Maybe putting more brakes on all planes? I've heard some planes have drag shoots on them, shouldn't all of them? Well, the solution most widely utilized is an EMAS.

EMAS is a series of boxes partially filled with a concrete powder-like material. No, not kidding this time. When a runaway aircraft crashes into these boxes, it slowly brings the aircraft to a stop. It essentially is a ground-based arrester system. This greatly reduces the chance of major damage to the aircraft or injury to its occupants. The EMAS of course is naturally dependent upon the size, weight, and speed of the aircraft. It is designed to accommodate most aircraft which operate out of the airport it is fitted for, at entry speeds into the EMAS at the end of the runway of 70 knots or less.

To date, there have been 15 incidents where EMAS systems have safely stopped 15 overrunning aircraft, carrying 406 crew and passengers, aboard those flights. EMAS is installed at 111 runway ends at 66 airports in the United States.

EMAS can be very expensive to install, anywhere from $2M up to $40M. Other than cost and maintenance, it has some other drawbacks as well. For whatever reason, the color of it is gray, the same color as most runways. While they do have yellow arrow chevrons on them, it is difficult to see them from the ground, and especially at night. I think you know where I'm going with this. If an aircraft inadvertently taxis onto one of these and sinks into it, that can be very expensive. Sure enough, this does happen from time to time. Many times, a crane is needed to extract the aircraft from the EMAS, and EMAS repairs can be costly.

I have been involved in a couple investigations wherein an aircraft inadvertently taxied into an EMAS. The first was at an airport in Massachusetts. The pilot of a small single-engine airplane simply thought the EMAS was the end of the runway. Once he was on it, the aircraft partially sunk into it. It was an expensive repair; however, the airport could not put complete fault onto the pilot as there was not a yellow line that leads the pilot off the runway at its end (lead-in line).

The second investigation I was on was in New Jersey. It involved a jet inadvertently taxiing into the EMAS at night. Unfortunately, it sank considerably into it. Cranes were needed to get it out. Since then, some airports have put red lights in front of the EMAS – further identifying it to pilots to avoid.

It always baffled me why these are not depicted better, such as red and yellow or red and white prominent markings on them. Alternatively, I am grateful for the lives and property they have saved.

The Cirrus Life

While there are many aircraft manufacturers, such as Piper, Cessna, Boeing, Airbus, there is one that is a bit more unique than others. This aircraft is a Cirrus. It's a single-engine piston-powered propeller-driven plane with four seats. Very similar in that regard to other single-engine propeller airplanes. However, each one comes with a parachute for the entire airplane. It is literally built into the airplane. There is a handle in the cockpit that in the unlikely event the aircraft cannot land safely, the handle can be pulled which will deploy the chute and the entire aircraft will settle to the ground instead of crashing into the ground.

To date, there have been over 100 chute deployments saving hundreds, if not thousands of lives. Some pilots ventured into weather that they were unable to manage, some came too close to rising terrain, and other situations that may have otherwise led them to disaster. I have investigated a few of these, two of which stand out more than the others.

The first was a pilot who took off in the northeast and soon after takeoff passed out. He eventually came to and was so disoriented, he elected to pull the chute. When this happens, the aircraft floats under the chute toward the ground at 70 ft./s. The landing gear is designed to absorb the majority of the impact and break off. This prevents the occupants from absorbing the full impact of the aircraft hitting the ground. This is most effective over the ground; however, this aircraft came to rest in a small lake. This resulted in a hard impact which caused back injuries to the pilot. He, therefore, received a CT scan in the hospital which further revealed he had a brain tumor he did not know about. This explains why he passed out. Who knows when, or even if, he would have ever learned he had this – had it not been for this accident.

The other investigation I conducted involved a pilot with nobody else on board flying from the northeast to the southeast before pulling the chute before coming to rest gently in a tree. Upon initial investigation, there was little to go on. Only the pilot was on board the airplane.

The prop did not exhibit any damage, which further led me to believe the aircraft was not producing power when the chute was deployed. In other words, the aircraft ran out of gas. After performing some fuel burn calculations, it made sense the aircraft ran out of fuel where it did. But this certainly was not the end of the investigation; it actually was barely the beginning.

There were a couple other factors that begged for further inquiry. First, this happened in the middle of the night, around 3 a.m. Most single-engine pilots flying alone do not fly deep into the night and early morning. It can be dangerous if you factor in sleep patterns, weather that cannot easily be avoided, terrain that cannot easily be seen, etc.

Also, the pilot left the scene of the accident and was not readily available for a report or a statement. Almost all pilots are very cooperative after an accident. This one was not.

While we can assume the aircraft ran out of gas, we must confirm that there were no other abnormalities with the aircraft. In other words, we must make sure nothing was wrong with the plane. Therefore, the post-accident wreckage inspection was performed. During this stage of the investigation, the engine, aircraft, flight controls, avionics, and anything else with regards to the performance of flight is examined and tested. As suspected, there were no abnormalities found with the aircraft, other than it having little to no fuel. But, upon further inspection we found cigarette butts on the floor of the cockpit. While this may be common practice in the 1960s and 70s, this was a first for me. Why would a pilot be smoking cigarettes in an unpressurized aircraft with no windows that can be opened? We further noticed burn marks on the dashboard of the airplane near the instruments. So, was the pilot smoking cigarettes during the flight and putting them out on the dashboard/panel? This made no sense whatsoever. What was further baffling our investigation is the pilot was not cooperative in answering these many questions we had.

Upon further searching of the aircraft, a couple syringes were found. As you might have guessed, the residue within the syringes was tested by local law enforcement and came back positive for heroin. The pilot never admitted that they were his and local law enforcement could never charge him with it. However, this certainly further explained the odd behavior of a pilot flying in the dead of night while smoking cigarettes and putting them out on the dashboard (instrument panel) until the plane ran out of gas, at which time the chute was deployed.

I later heard the pilot was a national auto parts dealer and eventually went to rehab. I am thankful he is alive and pray he is clean and sober.

Bang Bang

Believe it or not, airplanes are a magnet for bullets. Yes, you read that right. Bullets! I have participated in many investigations with aircraft damaged by bullets. How does this happen? Where do they come from? Are there aircraft that are targeted or is just a freak accident scenario? Well, the answer is yes.

One of my first investigations into bullets damaging an aircraft was a pilot flying a single-engine aircraft and his aileron, which is the part on the wing that turns the airplane, was hit by a stray bullet. He was flying pretty low and likely a hunter got bored and took a shot at him. This happened in the Midwest and thankfully the damage was pretty minor. After some re-skinning and painting, the aircraft was as good as new.

The next was an aircraft parked in California and the engine was shot. A bullet projectile was actually found inside the engine. While we never knew for sure, local gang activity likely spilled into the streets near the airport and a stray bullet struck a parked aircraft. A similar occurrence happened in Arkansas during deer hunting season. Can't make this up.

Last, but not least, was a helicopter that was shot in the tail section while inside a hangar. The pilot was actually in the hangar when this occurred, so he heard the bullet penetrate the hangar wall and the aircraft. After he looked around for a bit, he quickly realized what happened and what direction it came from.

There was a nearby house in the general direction where the bullet came from. He knew of the owner of the house. So, he contacted the owner and asked if he had been shooting off any rifles lately. Surprisingly, the answer was yes. The owner had just purchased an AR-15 and he was shooting it that very night. The pilot explained what happened and while the homeowner did not immediately accept responsibility, he did admit that it was possible. Little did the homeowner know, but the pilot recorded the conversation and submitted it to the police. The homeowner then agreed to pay for the damage and did not face any charges.

I have been around guns my whole life and even shot competition in college. In fact, I was invited to an event with other pilots and aircraft accident investigators for a round of trapshooting, which involves shooting moving targets with a 12-gauge shotgun. While I had never shot trap before, I had shot skeet which was a little similar. In skeet, the targets come from one of two towers. In trap, the targets come from ground-based spring-loaded launchers in multiple directions. A buddy of mine on my team simply told me to cover up the target with the end of the shotgun and keep the gun moving once you shoot it. It's kind of like a follow-through in a tennis or golf swing. So, that's what I did, and lo and behold I won the tournament. Not bad for a first timer!

Shooting is a great sport and can be a lot of fun. However, I was always taught that once you pull the trigger the bullet must end up somewhere. I wish everyone was cognizant of that. Just today as I write this, I received another report of aircraft damage from a gunshot.

Virginia

There are some accidents that will always stick with you. This is one of them. It has honestly taken me several weeks to start this chapter. This is a rough one. But here we go.

My office received reports of a small plane that crashed in a rural neighborhood in Virginia that resulted in two fatalities. Immediately when we hear this, we think husband and wife – or similar. We quickly learned this was not the case.

Being a commercial pilot based in New York, I had access to a fleet of small aircraft I could use to respond to an accident. So, one of the senior investigators and I flew a plane almost immediately down to Virginia to investigate this loss. We learned a little bit more about it as we were getting ready to roll.

On the day of the accident, the weather was really good. Some neighbors had got together in Virginia for a barbecue/cookout. One of the neighborhood attendees was a local pilot. He was telling stories about flying and one family was particularly interested in it. Their 12-year-old son really paid close attention to the amazing flying stories. Such a freedom. Such a view. Such an experience. He really wanted to go, and his parents absolutely supported it. Who would not want their son to experience such majesty? To broaden their horizons. To learn about things about flying like science, math, physics, theory. So it was then decided the neighbor pilot would take the neighbor's son flying with him. What an adventure!

Growing up I never had an opportunity presented to me like this. Many Sunday mornings my dad would take me to the airport to watch the Concorde land. It was incredible. He would also take me to airshows – which I loved with a passion (and still do!). One time we also saw a hot air balloon at our local church. Such grace in the sky. But never was I offered a flight in a small airplane. If it was available, he would have wanted it for me.

The pilot was later described as fairly meticulous and thorough.

However, he decided that as an added thrill he would buzz the neighborhood at a low altitude. Unfortunately, at one point he got so low he made contact with a tree which caused the aircraft to dip and strike another tree, ejecting both occupants out of their seatbelts.

The pilot was immediately killed. The passenger suffered major facial injuries. The parents rushed to his aid; however, his heart stopped before the paramedics arrived.

As aircraft accident investigators we must look at all the evidence available. This includes both physical evidence and witness statements. The physical evidence showed a strike to a tall tree, followed by a major impact to a lower tree. The rapid deceleration was so intense that both aircraft occupants were ripped out of their seatbelts and ejected from the aircraft.

Also, as part of the investigation we interviewed the parents of the deceased boy as they witnessed the accident. After describing the accident facts to us, I asked the parents if they had any questions. The father asked me a question I will never forget. It honestly still haunts me to this day. He simply said to me, "Tomorrow I must bury my son. How am I going to do this?"

I wish I could write more in this chapter, but I just can't.

Race Car, Anyone?

This particular crash has so many things wrong with it, so many that it is hard to find where to start. Let's first look at the background of the aircraft and owner. Many people may not realize how much money really is in car racing. And I mean professional car racing. Let's just say, it is a very lucrative sport.

Most well-known professional racecar drivers have their own aircraft. Many have their own fleet of aircraft. Some drivers are also pilots of their own aircraft. This accident involved an aircraft widely used in this sport. More on that later.

When people find out what I really do for a living, most are fascinated. Some, however, are shocked and a bit terrified. Most people say to me that I have a very interesting job and I must love it. I often respond telling them that is certainly true. Another question I get is if my job is difficult at times. The answer is definitely yes. It is especially hard when innocent passengers, especially children, fall victim to a plane crash. It is even harder when they are an innocent bystander on the ground. This particular accident involves both.

The July day was completely clear. Barely a cloud in the sky. School is out for summer break. The twin-engine plane took off from Daytona Beach and headed for Lakeland in Florida. Should have been a short 45-minute flight for a lunch meeting. The pilot and passenger settled in for the quick journey. But something went terribly wrong.

But what had gone awry? What happened? All we knew in the beginning of the investigation is the pilot declared an emergency only 10 minutes into the flight at a cruise altitude of 6,000 feet. His last transmission occurred approximately one minute before striking two houses.

So the investigation continues. We know that the weather was clear the morning of the accident. So, we can likely rule out any weather-related factors causing the accident. The aircraft crashed and burned, likely showing that it was properly fueled. We also know from fuel records that the aircraft was full of fuel upon takeoff and was only 10 minutes into the flight. Thus, fuel starvation or exhaustion is likely not a factor of contributing to the accident.

Remember, as accident investigators, we investigate people, machine, and the environment. The pilot was appropriately rated and current for this aircraft type. We already know that weather was likely not a factor. Ground evidence and witness statements suggest that both engines were producing proper power upon impact.

Let's investigate that last transmission. After declaring an emergency, the pilot said he would be shutting down all electrical equipment, including the radios. Why would the pilot do that? What was he trying to do?

After speaking with the maintenance personnel who usually worked on the airplane, we quickly learned that the aircraft experienced a mild smoke condition the day before. However, instead of the aircraft being grounded for extended maintenance inspections, it simply flew again the next day.

When doing a fire investigation with an aircraft, we must discern what is burned as the result of an in-flight fire versus a post-crash fire. Certain parts of the aircraft tend to burn in the air if on fire. In this case, unique amounts of soot were found around the aircraft wiring, particularly around the avionics. It appears based on all these factors, including the ones we ruled out, the most likely cause of the crash was loss of control during an in-flight fire. The cabin likely filled up with smoke and the pilot/passenger was just along for the ride at that point.

As mentioned before, the aircraft struck two houses on the way down. In the first house, a mother and her six-month-old son could not escape the smoke. In the second house, a four-year-old girl died from the impact. Her 10-year-old brother jumped out of a second-story window and survived. As did the parents. All received considerable burns.

It was opined that the aircraft should have never flown on the day of the accident knowing that just the day before it had a smoke event that was not properly investigated. It was also alleged that this flight was simply a joyride and was not a work-related or mission trip. As a pilot myself, I recently experienced the smell of smoke shortly after takeoff combined with some heat emanating from the forward floor of the cockpit. Within

60 seconds I had the plane back on the ground. The maintenance crew inspected it the next day and sure enough, a tube had come loose after a clamp failed, causing it to rest on the hot engine and melt. Had the aircraft flown again without maintenance inspecting it, the result could have been catastrophic.

Back to this accident. In the end, the two aircraft occupants and three bystanders on the ground lost their lives, including two young children. When people ask me if my job is difficult at times, this accident reminds me that the answer is yes.

Pilots and Brazil Don't Mix

On September 29, 2006, Brazil experienced its worst air disaster to date when Gol Linhas Aereas Inteligentes Flight 1907, carrying 149 passengers and six crew members, crashed nose-first into the dense Amazon rainforest of Mato Grosso state after a midair collision with a private Embraer Legacy business jet. There were no survivors of the 737.However, both pilots of the Embraer business jet did survive.

Radar contact was lost with Gol's Boeing 737-800 during the flight from Amazon's principal city of Manaus to Brasilia, the nation's capital. An air force rescue team located the wreckage the following day and began an unsuccessful search for survivors.

The smaller Legacy jet, enroute from the Brazilian factory (of Embraer), successfully negotiated an emergency landing at an air force base in Serra do Cachimbo, in spite of extensive damage to the plane's wingtip and tail.

—◇—

Brazilian and U.S. investigators investigated voice and data recorders from both aircraft. At the request of prosecutors, the judge seized the Legacy pilots' passports to ensure they remained in the country for questioning in order to ascertain the cause of the collision. They were not allowed to leave Brazil while the investigation is ongoing. Prosecutors said they could charge the American pilots with involuntary manslaughter if they were found responsible for turning off the transponder, which is illegal under Brazilian law. They were eventually released.

The Legacy executive jet did not comply with an order by the control tower to descend to a lower altitude. The Legacy was flying at 37,000 feet and did not descend to 36,000 feet as requested. The Gol aircraft was also flying at 37,000 feet in the opposite direction.

A chief concern was the failure of the Traffic Collision Avoidance System (TCAS) to notify either cockpit of the other aircraft's altitude and proximity. Authorities also were trying to determine why the westbound Legacy jet was flying at an altitude reserved for eastbound air traffic. When an aircraft flies east, it should be at an odd-numbered altitude, such as 35,000 feet. When an aircraft flies west, it should be at an even number, such as 38,000 feet.

Investigators examined voice and data recorders recovered from the jetliner, but the National Civil Aviation Agency said one of the voice recorders was missing data. The unit is essential for analysis and the military was searching for the missing parts. The NTSB (National Transportation Safety Board), the FAA (Federal Aviation Administration) along with Embraer and Boeing assisted in the investigation.

The evidence collected during this investigation strongly supports the conclusion that this accident was caused by the corporate jet and GLO1907 following ATC clearances which directed them to operate in opposite directions on the same airway at the same altitude resulting in a mid-air collision.

The loss of effective air traffic control was not the result of a single error, but of a combination of numerous individual and institutional ATC factors, which reflected systemic shortcomings in emphasis on positive air traffic control concepts.

Contributing to this accident was the undetected loss of functionality of the airborne collision avoidance system technology because of the inadvertent inactivation of the transponder on board the corporate aircraft.

Further contributing to the accident was inadequate communication between ATC and the private jet's flight crew.

Hangar Foam for All

What could this chapter possibly be about? What in the hell is hangar foam? Is it dangerous? What is it used for? If you are thinking that hangar foam is used to protect airplanes from bumping into each other, you are dead wrong (although I wish there was something like that – which would make my life much easier).

Hangar foam is a form of fire suppressant used for two reasons. The first is obviously to put out fires in an airplane hangar. The second is to prevent aircraft from catching fire once a fire is detected. There is one issue with this. Off the top of my head, I cannot think of one time there was a hangar fire that was put out by a hangar foam suppression system. That's right. Just about every time a foam fire suppression system has gone off/activated in a hangar, it has not been for a fire.

So why does this happen? Why does this fire suppression system activate when there is not a fire? There are many theories, but here are a few. Some believe water moisture condenses in the control panel and creates a circuit that activates the system. In one particular instance, a nearby water heater blew and actual water penetrated the control panel which activated the system. In another instance, somebody drove a golf cart into the control panel which closed the circuit creating the activation.

When the system is activated, large cannons shoot out a huge amount of foam in a short period of time. Yes, it can suffocate someone if they are in the wrong place at the wrong time. Also, many types of this foam are considered corrosive and dangerous. They can cause horrific damage to aircraft as well, both inside and out. The damage from one of these systems can often exceed the cost of damage from an actual plane crash.

This chapter is in the book to remind people how dangerous and dynamic many parts of aviation can be, not just from plane crashes. Many people have been hurt or killed by planes on the ground. Tires and oxygen bottles can explode if not properly handled. People can get sucked into a jet engine running on the ground. People have even been severely hurt or killed just from ground service vehicles for aircraft support.

It is important to realize that safety is paramount no matter what part of aviation you are involved in. It really can be dangerous.

The Perils of Being an Airport Security Employee

There are so many jobs someone can have at an airport. Especially at the big airports. Many describe it as a full-blown city. In many ways, in most ways, it is. Its own police department, its own fire department, elected officials, managers, support staff, etc.

One of the major roles at an airport is cargo handling. And no, that does not mean passenger luggage. A major airport can move a lot of cargo. Not just with UPS, FedEx, DHL, and alike, but major shipping cans of goods such as clothing, large pallets with items like flowers, and even seafood.

Well, cargo needs to be guarded and therein is the need for security employees. People may think this is a mundane job. But it is quite involved. There is inbound cargo to be inspected, outbound cargo that must match the manifest, security for cargo while in storage (both for cold and normal temperature storage facilities), personnel control, equipment management, and many other duties.

Often cargo will end up on a large dolly system being towed by a large tug. The flatbed dolly that the cargo sits on often has wheels to easily move cargo on and off it. You might not realize this, but this part is an intricate portion of the cargo service at an airport.

Early one morning before the sun came up, a cargo employee working security was checking the paperwork on top of cargo containers that were on a flatbed dolly that was to connect to a tug for transfer. The tug driver started going forward. The security guard slipped, and his foot went under a wheel of the flatbed dolly. Since it was early in the morning, the wheels were still wet. This was from dew, kind of like in the morning grass. Most of the skin from his foot detached, creating a degloving injury. While he eventually recovered, he will never be 100% again.

So we have established the cargo cans and pallets usually end up on flatbed dollies with rolling wheels. Cargo is loaded and offloaded from these during short airport transfers, usually from an aircraft to storage. Once loaded onto the flatbed, how are the containers and pallets secured during transport? In most instances, they are strapped on.

Generally, a cargo worker will attach a strap and latch to one side of the flatbed dolly, then toss it over the containers to the other side before securing it. This usually works out well unless a security guard is on the other side. It may be especially hard to notice this at 3 a.m.

Sure enough, one morning around 3 a.m. a cargo worker tossed over a strap with a large metal latch on it over the cargo containers and struck a security employee in the head, knocking him out. To add insult to injury (literally), the security guard injured his shoulder and hit his head again upon landing on the pavement. While he eventually recovered the best he could, he did claim a traumatic brain injury with lasting effects.

These are things that most people don't think about when considering the dangers of an airport. But like any major city, there are many hazards, which can and do exist – and must be mitigated as much as possible.

Look Out Below!

Pilots and doors can be a bad combination.

In many ways, pilots have a lot of duties not directly related to flying. They must ensure the aircraft is properly loaded, weight is properly distributed, fuel is properly pumped into the aircraft, catering is on board, luggage is on board, etc. However, in many ways, pilots do not have a lot of duties. There is a ground crew that loads luggage, fuel, catering, and many other things that go on and off the airplane. While the pilot is ultimately responsible for everything, most ground duties do not directly include the pilot.

Basically, so far this chapter is an oxymoron. The pilot has many duties, and the pilot has few duties. You will understand this further as we delve more into this chapter.

The story starts out like this. It was an early morning flight with the aircraft parked at the gate as the sun was just coming up. Passengers were not on board yet, but the crew was. Evidently, the pilot was not happy with the speed of the provisions being loaded onto the aircraft. He went to go check it out.

When a provision truck goes up to an aircraft to load it, generally there is a plate that is placed between the truck and the airplane to create a bridge between the two vehicles. That way, the truck does not have to get too close to the plane and risk hitting it. You can also roll carts across it with ease.

This particular pilot was in a rush and feisty that morning. He decided to walk onto the provisions truck over the bridge plate and inspect the loading progress. His stride was fast and steady. On the way back to the aircraft from the truck, his stride was unchanged. He was so fast in fact that he caused the bridge plate to slip, and he fell over 20 feet to the ground below him. Yes, he slipped between the aircraft and truck.

While he survived the fall, he had major injuries to say the least. In addition to some fractures, he also suffered a traumatic brain injury that would render him unable to perform his flying duties ever again. He also would require some level of care for the rest of his life.

On another flight, maintenance was called out to run some tests on the avionics in the cockpit. This was being done in the middle of the night before passengers were loaded. Not even the flight attendants were on board yet. When the pilot left the cockpit, he proceeded to the airstairs, which some aircraft use instead of a jet bridge. There was a small gap between the airstairs and the aircraft, which the pilot fell through. He was airlifted to a local hospital with multiple fractures and internal organ injuries. When he was prepped for surgery, he started shaking violently. It was then determined he was an alcoholic and had to be fed beers to calm him down for surgery. Between this and the injury, his career was over.

Airports and aircraft can be very dangerous at times. While both pilots lived, their lives will be forever changed.

Block Island, You Blockhead

When most people think of Rhode Island, they hardly ever imagine an island south of it in the Atlantic. Well, Block Island is just that. While it's a lot like Newport, and so much so it has 17 miles of beaches, beautiful homes, bluffs, it is quite small with only 1,000 active residents. By the way, some of them have never left the island. Situated between Montauk and Martha's Vineyard, it truly is a majestic place.

Block Island even has its own airport. It's a bit small with one runway that is about 2,500 feet long. Now I know what you must be thinking. It's a tricky runway with water on all sides sloping down a cliff. That is not the case at all. It's a simple runway and a very simple airport. However, I have been there on business many times if you know what I mean.

The airport opened in 1950 and it is over 100 feet above sea level. It does not have a control tower, so pilots simply announce themselves while flying in the pattern over a common radio frequency. In other words, pilots broadcast their position on the runway so other pilots can avoid them.

On one instance I was there doing an investigation over a holiday weekend. A plane experienced a wingtip strike on the landing which caused it to cartwheel into a row of parked airplanes. Thankfully, there were no major injuries as most of the energy of the landing aircraft had bled off by the time it struck parked airplanes.

While this accident was unique in nature, I could not help but also notice there were several other damaged aircraft at the airport that I was unaware of. I saw a corporate aircraft with a gash in the tail. When I asked what happened, I later found out a nearby helicopter hit it with its running rotor blade.

On the other side of the airport, I saw a six-seater prop airplane with the prop bent to hell and the belly of the airplane scratched up badly. The local airport representative told me the aircraft went off the end of the runway and through the perimeter fence.

So literally, there were three accidents at this airport over a single holiday weekend! I can't make this up.

Keeping in mind that while Block Island is a beautiful place to visit, bad things can happen. And it's interesting to note that bad things sometimes happen in threes.

Flying Wild Alaska

One of the most amazing places in the world to go flying is in Alaska. I have never seen so many seaplanes in one place. The combination of land and water utilized for flight is just unimaginable. Alaska really has it all. Large modern cities. Nearly untouched backcountry. Mountains, rivers, glaciers, amazing wildlife, and abundant natural resources. While most areas can be traveled to by car, boat, or train, some are only accessible by aircraft.

God truly made a work of art when he created Alaska. If you are the type of person who really connects with natural beauty, wild animals, and nature, then Alaska is for you. However, as incredible as Alaska is, it's unfortunately a haven for plane crashes. It seems like once a year there is a tour group airplane or helicopter that hits the side of a mountain in marginal to poor weather.

There is one particular accident that stands out in my mind. A boyfriend and girlfriend were both pilots with the same cargo company that brings goods to and from small towns in Alaska. They each flew a medium-sized cargo aircraft that only required one pilot. After doing this for a few seasons, they both ended up in the same small town making deliveries and decided to fly back to base in loose formation. With each of them flying an aircraft single pilot, the two aircraft flight back to base was fairly uneventful until the boyfriend thought it would be a good idea to fly a circle around his girlfriend's aircraft. While she was initially against the idea, he elected to do it anyway.

The boyfriend circled his airplane around the bottom of his girlfriend's airplane. As he circled toward the top of his girlfriend's airplane, both aircraft encountered each other creating a mid-air collision. The propeller from the girlfriend's airplane chopped up the steering control of the boyfriend's airplane, sending it straight to the ground. While the girlfriend's airplane was significantly damaged, she managed to limp her airplane back to base without incident.

She later found out her boyfriend perished in the accident. She was so shaken from this event she never flew again. The part of Alaska the aircraft went down in was so remote we had to wait until spring to dig it out of the ice and snow.

While this accident is extremely tragic, it serves as a stern reminder that one simple mistake, one lapse in judgment, can easily lead to tragedy. I can't imagine how the girlfriend must feel and must go on living with. The people I am close to I care for dearly and I am very protective over. This reminds me how precious that is and how quickly it can be lost.

Upstate New York

As I start writing this final chapter, unrelated, I am reminded that just last month two pilots I knew died in a tragic plane crash. It was recently learned that the parking brake was left in the engaged position in their small jet while attempting to takeoff from a small runway. The engines produced full power, but it was just not enough to combat the brakes being in the locked position. Their jet went off the runway and straight into a building. There were no survivors.

OK, back to the story of this chapter starting in upstate New York. It was during the rainy season in the Northeast as late spring started turning into summer. An aircraft encountered in-flight hail and was substantially damaged but managed to land at an airport in upstate New York.

A local pilot friend of mine and I decided to hop in our plane to check it out. Sure enough, while on the way up we noticed a storm was fast approaching the airport we intended to land at. Yes, this can be typical for this time of year, but still never is welcomed from up in the air. We did some quick time and speed calculations, then determined we could land before the storm would hit the airport we were intending to land at.

As we landed, we could see the storm just off in the distance. We got the plane into a parking spot and the storm hit right then and there. As we stayed in the airplane to let the storm quickly pass, we noticed a vehicle driving up and down the runway during the storm. It was a newer sports car and was traveling down the wet runway at tremendous speeds. The airport we were at did not have a control tower, like Block Island. Thus, this might not have been the smartest idea for someone to be driving at high speeds down an active runway, storm or not.

Sure enough, the driver of the sports car ended up being the owner of the airplane with hail damage we were traveling to investigate. The airplane owner was extremely friendly and said he was testing out his new sports car on the runway to see if the antilock braking worked well. Might not have been the very best decision to make in front of the investigators here to review the damage to your airplane.

Coincidentally, the investigation went very well, and the aircraft owner was very helpful. Before we met him, we obviously had concerns that he would be an arrogant know-it-all. Thankfully, that was not the case. He had a very successful career that he told us about, and you could see his zest for life. I ended up liking him a lot, despite his initial reckless actions. His aircraft was eventually repaired, and we all continued our lives.

A few years later around the same area in upstate New York, a private small fighter jet crashed into a snowbank, killing the pilot. I received a call a couple days later from the pilot I flew to that area with. He said I knew the deceased pilot of the local crash. I told him that I doubted it I did. Unfortunately, he was right. It was the pilot we had previously met at that nearby airport several years ago.

Apparently, he had recently purchased a small private fighter jet, and this was his inaugural flight home. His family and friends gathered at the airport where they met him to watch his inbound landing. Instead of doing a normal approach and landing, he pulled the plane straight up and aerodynamically stalled, which resulted in impact with terrain as he did not have enough altitude to recover. In other words, he tried to show off for everyone and did not make it. I was then reminded that some people's zest for life can overbear rational thoughts and actions. This accident was tragic as it was witnessed by his friends and family, and he left behind a wife and kids. If anything good can come out of this, maybe it can serve as a reminder to be careful when pushing limits.

Conclusion

I have had the honor and privilege of being an aircraft accident investigator for over two decades. Much of what I have learned I have put into safety presentations to show other pilots that smart people can sometimes make bad decisions. If one life can be saved from learning from others, then it was worth it! Thank you for letting me share some of my more memorable experiences with you in this book.

Made in the USA
Coppell, TX
19 March 2026

74193230R00049